Housing
228

水中的食肉植物
Carnivores in the Water

Gunter Pauli

[比] 冈特·鲍利 著

[哥伦] 凯瑟琳娜·巴赫 绘

李原原 译

上海远东出版社

丛书编委会

主　任：贾　峰

副主任：何家振　闫世东　郑立明

委　员：李原原　祝真旭　牛玲娟　梁雅丽　任泽林

　　　　王　岢　陈　卫　郑循如　吴建民　彭　勇

　　　　王梦雨　戴　虹　靳增江　孟　蝶　崔晓晓

特别感谢以下热心人士对童书工作的支持：

匡志强　方　芳　宋小华　解　东　厉　云　李　婧

刘　丹　熊彩虹　罗淑怡　旷　婉　杨　荣　刘学振

何圣霖　王必斗　潘林平　熊志强　廖清州　谭燕宁

王　征　白　纯　张林霞　寿颖慧　罗　佳　傅　俊

胡海朋　白永喆　韦小宏　李　杰　欧　亮

目录

水中的食肉植物   4

你知道吗?   22

想一想   26

自己动手!   27

学科知识   28

情感智慧   29

艺术   29

思维拓展   30

动手能力   30

故事灵感来自   31

Contents

Carnivores in the Water   4

Did You Know?   22

Think about It   26

Do It Yourself!   27

Academic Knowledge   28

Emotional Intelligence   29

The Arts   29

Systems: Making the Connections   30

Capacity to Implement   30

This Fable Is Inspired by   31

一棵水车草因为看到一条蛇在水里游动而感到惊讶。

"你是不是走错地方了？"水车草问。

"走错地方？"海蛇回答道："这是我的家。"

A waterwheel plant is surprised to see a snake swimming in the water.

"Are you not in the wrong place?" the waterwheel asks.

"Wrong place?" the sea snake replies. "This is my home."

水车草感到惊讶……

A waterwheel plant is surprised ...

海洋采矿……

All this sea mining …

"我一直以为蛇应该生活在森林或沙漠，而不是水里。那你是什么时候到这儿的？"

"我不知道自己来这儿多久了，也不知道还能在这儿待多久，毕竟这儿的污染如此严重。海洋采矿，还有那些噪音，都使我们的生活变得非常困难。"

"I always imagined snakes living in a forest, or the desert, but not in the water. When did you get here?"

"I have no idea how long ago I arrived. And I don't know how long I will still be here, with all this pollution. All this sea mining, with all the noise, is making life very difficult for us."

"告诉我，如果你连潜水装备都没有，你怎么呼吸？"水车草问道。

　　"潜水装备！真有趣。"海蛇笑了。"只有人类需要瓶装空气。除了我的肺，我还可以通过皮肤呼吸，氧气直接进入我的大脑。我的鼻孔上还有瓣膜，可以防止水进入肺部。"

"And tell me, how do you breathe if you don't even have any scuba gear?" Waterwheel asks.

"Scuba gear! Very funny," Sea Snake laughs. "Only people need bottled air. I don't, as apart from my lungs, I can also breathe through my skin, with the oxygen going straight to my brain. And I have valves over my nostrils to stop water getting into my lungs."

......氧气直接进入我的大脑。

... oxygen going straight to my brain.

……适应了这里的新生活。

... adjusted to a new life here.

"问题就这样解决了！你跳进水里，适应了这里的新生活。"

　　"适应新的现实？这不就是我们在自然界做的吗？毕竟，我们唯一可以肯定的是，一切都在不断变化……"

　　"蛇，你太聪明了，而且你如此擅长观察周围发生的事情。你的小眼睛很强大吗？"

"Problem solved, just like that! You got into the water, and adjusted to a new life here."

"Is that not what we all do in Nature? Adapt to new realities? After all, the only certainty we have, is that everything changes all the time…"

"You are so wise, Snake. And so observant of what goes on around you. Are your small eyes very strong?"

"不，我的眼睛不是那么有用。但我的脑袋里确实有传感器。我们在水中能感觉到振动，而在陆地上，我们必须触摸物体才能感受到它们。"

"你不仅聪明和善于观察，你看起来也很高科技！"

"你可以这样看，但我们不是都有独特之处吗？这就是多样性的力量。"

"No, my eyes are not all that useful. But I do have sensors in my head. We feel vibrations in the water, whereas on land, we have to touch things to feel them."

"You are not only wise and observant, you seem to be high tech too!"

"You could look at it that way, but don't we all have something unique? And therein lies the power of diversity."

我的脑袋里有传感器。

I do have sensors in my head.

你知道维纳斯捕蝇草吗？

you know the Venus flytrap?

"我发现，人类往往不能很好地理解多样性。他们喜欢把所有人和所有的东西都放在一个盒子里，你只能做别人要求你做的事……不能有例外。"

"我能问一下你在做什么不该做的事吗？"

"你知道维纳斯捕蝇草吗？"

"I find that diversity is often not well understood. People like to put everyone and everything into a box, and you must only do what is expected of you… Exceptions are not appreciated."

"May I ask what it is you are doing that you are not expected to do?"

"You know the Venus flytrap?"

"你是说那种以飞虫为食的食肉植物？"蛇问道。

"嗯嗯，是的。我就是水中的捕蝇草。"

"可是你怎么能在水里捉住苍蝇呢？水的密度这么大，一棵植物不可能在水中快速运动从而捉住任何东西！"

"You mean that carnivorous plant that eats flying insects?" Snake asks.

"Yes. Well, I am the water version."

"But how can you catch flies in the water? With the water so dense, a plant cannot possibly move fast enough to catch anything!"

我是水中的捕蝇草。

I am the water version.

……而是虾和蝌蚪。

... but shrimp and tadpoles.

"我抓的不是苍蝇，而是虾和蝌蚪。还有很多幼虫，我会在他们变成会飞的成虫之前捕捉他们！而且，我在水里捉虫的速度比捕蝇草在陆地上还快。"

"但是你没有任何肌肉，你是怎么一把捉住猎物的呢？"海蛇问。

"靠细胞的力量！我的捕虫夹关闭的速度比人类眨眼的速度快十倍。"

"I don't catch flies, but shrimp and tadpoles. And a lot of larvae, before they turn into flying insects! What's more, I am faster in the water than the Venus flytrap is on land."

"But you don't have any muscles, so how do you snatch your prey?" Sea Snake asks.

"Cell power! I can shut my trap ten times faster than people can blink their eyes."

"震惊！难怪他们也叫你陷阱。水中有足够的食物让你们水车草生存下去吗？"

"以前没有问题。但随着湿地里新建筑的出现，情况发生了变化……幸运的是，我在加州海岸找到了新家。但又不幸的是，在这儿我被称为外来者。"

"我只希望他们不要认为你是非法移民！"

……这仅仅是开始！……

"Astounding! No wonder they also call you the snap-trap. And is there enough food in the water for you snap-traps to survive?"

"There used to be, at home. But it changed with all the new buildings in wetlands... Fortunately I have found a new home here along the coast of California. But unfortunately, here I'm called an alien."

"I just hope they do not consider you an illegal alien!"

... AND IT HAS ONLY JUST BEGUN!...

……这仅仅是开始！……

... AND IT HAS ONLY JUST BEGUN! ...

Did You Know?

你知道吗?

6 000 000 年

Six million years ago snakes slithered into the sea. Descendants developed flattened paddle tails, and the ability to breathe through the skin. Sea snakes outnumber the sea turtle species by ten to one.

600万年前，蛇类蜿蜒滑入海里。它们的后代进化出了扁平的桨状尾和通过皮肤呼吸的能力。海蛇的数量是海龟的10倍。

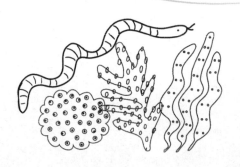

There are 70 species of sea snakes that spend their entire life in the water. Their preferred habitats are coral reefs and meadows of sea grass. Their population is declining rapidly as a result of climate change, pollution, overfishing and mining.

有70种海蛇一生都生活在水里。它们最喜欢的栖息地是珊瑚礁和海草甸。由于气候变化、污染、过度捕捞和采矿，它们的数量正在迅速减少。

地震测试依赖水下巨大的爆炸声来探测石油和天然气，这种测试会损害海蛇的听力和它们探测振动的能力。地震测试破坏了海蛇在野外狩猎和躲藏的能力，导致它们的数量大幅减少。

Seismic testing, which relies on loud underwater blasts to explore for oil and gas damages sea snakes' hearing and their ability to detect vibrations. It disrupts their ability to hunt and hide in the wild, leading to the decimation of their numbers.

Sea snakes are highly specialised hunters, going after specific prey like damselfish eggs, frogfish, or gobies. As a result, ten or more sea snake species may live in the same habitat as each has its own prey species.

海蛇是高度专业化的捕猎者，捕食特定的猎物，如雀鲷卵、蟾鱼或虾虎鱼。这使得10种以上的海蛇可能生活在同一栖息地，因为每一种海蛇都有自己的猎物种类。

Over the past 150 years, 90 percent of the habitat of the waterwheel has vanished, in at least 32 of the 43 countries where it naturally occurs.

在过去的 150 年里，在 43 个水车植物自然生长的国家中，至少有 32 个国家，其境内 90% 的水车植物栖息地已经消失。

Waterwheels have survived for the past 50 million years by living in marginal habitats with limited nutrients, surviving on invertebrates. Warming climate and drying land for construction causes eutrophication, which leads to extinction.

在过去的 5 000 万年里，水车植物一直生活在营养有限的边缘生境，靠无脊椎动物生存。人类的建设使得气候变暖、土地干燥，导致富营养化，致使水车植物灭绝。

The waterwheel is a carnivorous, rootless, wetland plant living in swamp-like bodies of water. It leads a double life: of being an ecological menace in some parts of the world and an environmental victim elsewhere.

水车植物是一种食肉、无根、生活在沼泽中的湿地植物。它过着双重生活：在世界上某些地区是生态威胁，而在其他一些地区则是环境的受害者。

The large bumblebee is disappearing in the UK, while outcompeting native pollinators in Argentina. The Burmese python has gone extinct in its native environment, but has become a pest in Florida. The sea lamprey is overfished in Southern Europe, but is invasive in the Great Lakes of the United States.

英国的大黄蜂正在消失，与此同时，大黄蜂的入侵使得阿根廷的本土传粉昆虫逐渐消失。缅甸蟒蛇在其原生环境中已经灭绝，但在佛罗里达却成了一种有害生物。七鳃鳗在南欧被过度捕捞，但在美国的五大湖区则是入侵者。

How do you think it feels to be an illegal alien?

你认为作为一个非法移民会是什么感觉?

Could you feel at home in a place different to where you were born?

在一个与你的出生地不同的地方，你会有家的感觉吗?

Are people able to adapt to changes, like snakes do?

人类能像蛇一样适应变化吗?

What do you use, besides your eyes, to perceive?

除了你的眼睛，你还用什么来感知?

Species that are at risk of going extinct in one part of the world, while being considered a pest in another part of the world, is a sign of imbalance in our environment. Such species can serve as a source for the recovery of the species that has their survival threatened in their native environment. Make a list of all the species that are considered pests elsewhere, while threatened or endangered where they naturally occur. Share this information with others, and discuss ways in which the pest species can be used to ensure the survival of the species in their native environment.

有些物种在某些地方濒临灭绝，而在另一些地方却被视为有害生物，这是生态不平衡的表现。这些物种可以作为一种资源，以恢复那些在其原生环境中生存受到威胁的物种的数量。请你列出一些在其他地方被认为有害，同时又在原生环境受到威胁或濒临灭绝的物种。与他人分享这些信息并讨论如何利用这些有害物种，以确保该物种在其原生环境中生存下来。

学科知识
Academic Knowledge

生物学	水车植物捕虫夹关闭的速度比捕蝇草快10倍；由于是食肉植物，水车植物不会与其他水生植物争夺营养；海蛇是眼镜蛇科的一种毒蛇；皮肤的呼吸；一些海蛇的鼻子和头顶之间有一个血管密集的区域，氧气可以从水中直接输送到海蛇的大脑；有些海蛇上岸后会在树上、岩石缝隙或石灰岩洞穴中产卵；海蛇有像桨一样的尾巴，可以在水中推进。
化 学	水车植物只能生活在酸性的水中；它的捕虫夹中含有磷酸盐消化酶。
物 理	排水的同时，水车植物的捕虫夹在10毫秒内迅速关闭；通过改变叶片细胞内的压力来加速，从而实现捕虫夹的快速关闭；捕蝇草采用液压系统；海蛇的游动方式是沿着腹部形成"龙骨"，增加表面面积，并通过横向波动帮助推进；海蛇体内的传感器可以远距离感知振动。
工程学	由人类引起的振动，如摩托艇和地震调查，对海蛇数量有潜在影响。
经济学	企业的竞争力取决于其适应新环境的能力；对市场的适应性还取决于快速变革的能力，以及通过加速摊销来适应产品和流程的不连续性。
伦理学	发展农业时引进非本地物种，可能导致生物入侵；为根除水车植物而使用的除草剂，也会杀死其他水生植物。
历 史	海蛇在900万到2 000万年前从陆地迁徙到海洋；水车植物已经生存了5 000万年，但直到1696年才被植物学家发现；查尔斯·达尔文证明了水车植物是食肉植物；大约在公元1300年，波斯潜水者用龟壳制作原始的护目镜；1943年，埃米尔·加格南和雅克·库斯托共同发明了一种带有需求调节器的自主潜水装备。
地 理	水车植物分布范围从俄罗斯的亚北极区到澳大利亚南部，从非洲西部到澳大利亚东部；澳大利亚和南非金伯利北部是水车植物唯一的避难所；海蛇生活在加利福尼亚湾的珊瑚礁，以及印度洋和太平洋的温暖水域。
数 学	水的密度为1 000千克/立方米；空气的密度只有1.275千克/立方米；海平面上的水与山上的水密度不同，湿度的差异也会对水的密度产生影响，计算时需要调整。
生活方式	水车植物能反映湿地建设热潮所导致的淡水生态系统恶化情况。
社会学	社会对移民接受程度的变化；可预测性或把所有东西都"放在盒子里"的欲望，总是期待同样的结果。
心理学	被视为外来者的耻辱，更糟糕的是，被视为非法外来者；知道有一个安全的家的重要性。
系统论	水车植物生活在相互连接的湿地上，水鸟在这些水体之间传播种子和孢子，所以一个地区的退化会在其他地区产生连锁反应；三分之二的湿地遭到破坏，水车植物因而失去栖息地，18世纪记录的水车植物有400种，现在只剩下50种，其中75%生长在切尔诺贝利附近。

情感智慧
Emotional Intelligence

水车草

水车草认为蛇的出现是一种入侵。在她看来，蛇生活在水里是不正常的。她想知道蛇是如何在水中生存的。她对蛇轻松适应新环境的能力表示惊讶。她意识到有关多样性的问题之一是人们坚持认为一切都应该"符合常规"。她为自己比维纳斯捕蝇草捉虫速度更快而感到自豪。她在两种现实之间左右为难：一种是在她的原生环境中水车草濒临灭绝，另一种是她在现在生长的地方被视为外来者。

海　蛇

海蛇坚持认为自己属于大海。他分享了自己不确定的未来。海蛇告诉水车草他是如何通过皮肤呼吸，氧气是如何直接进入大脑的。当受到挑战时，他表现出洞察力，说明所有生物要么适应，要么死亡。他说话谦逊，有自知之明，知道自己的局限性。他通过询问水车草做了什么特别的事情来表达同情。他很想了解水车草是如何生活的。他制造了越来越多的共同话题，分享了他对水车草"移民"的担忧，并表达了他的希望，希望水车草不要被视为"非法移民"。

艺术
The Arts

是时候学习怎么绘制卡通片了！水车草捕虫夹关闭的速度如此之快，以致于人眼无法察觉。为了研究它，需要把动作记录在胶片上，然后用慢动作回放。我们需要画12幅画，每幅画都展示这个动作的一个渐进部分。保持背景和颜色相同。每一张图片都是一帧，用手机拍下这12帧中的每一帧，你就可以把所有的图像按照正确的顺序放在一起。把它们放在一个演示文稿中，你就会拥有自己的动画电影！

思维拓展
Systems: Making the Connections

全世界的环境保护工作都面临着多重挑战。随着自然栖息地的减少，成千上万的物种处于危险之中。虽然外来物种的引入经常被认为是对本地物种的一个主要威胁，但过去的这些错误可能会产生一些意想不到却令人欣喜的结果。水车植物显然是一种入侵物种。在世界各地，尤其是在澳大利亚，水车植物的栖息地正在迅速消失，但现在它们有机会在其他地方保持一个强大而充满活力的群落。幸运的是，水车植物并没有在与其他水生植物的竞争中胜出，而是与它们和谐相处。然而，生态保护学家并不总是欣赏水车植物对环境保护的贡献。入侵物种的话题可以被扩展至更广的范围。一些劳动力接收国通过确保那些离开祖国的人得到训练和获得专业经验，成功地发展了本国的经济。移民返回家园创办新企业，甚至成为新工业部门的先锋。这种"人才回流"现象，对中国台湾、新加坡和韩国经济的发展起到至关重要的作用。一个人与他出生、成长、生活的地方的联系在他的一生中都是很重要的。正如弗吉尼亚北部的水车植物能够为生态系统的恢复提供种子一样，通过成功地使受过教育和训练的回国移民重新融入社会，经济也能够恢复。这些背井离乡的人往往是最具进取心和勇气的人，而且他们是一种社会财富，因为他们是社会上最富有活力和足智多谋的人。

动手能力
Capacity to Implement

随着工业活动的增加，如铺设海底电缆或在海底采矿中使用水下炸药，水下噪音污染已成为一个问题。水是声音的良导体，所以你能想象水中的声音比空气中的声音更响亮或更柔和吗？使用水下麦克风进行声音实验。把它放在一个水缸里，听一听水下空气软管喷出的气泡的声音。比较一下你在没有麦克风的情况下听到的气泡的响度。研究一下如何利用物理学来设计系统，以减少水下噪音污染。这可能是一个长期目标，但要"跳出框框"，进行创新。

故事灵感来自
This Fable Is Inspired by

布兰奇·丹纳斯塔西
Blanche D'Anastasi

　　布兰奇·丹纳斯塔西出生在澳大利亚北昆士兰，2008 年毕业于昆士兰詹姆斯库克大学，获得理学学士学位。她目前正在攻读濒危海蛇基因组学的博士学位。布兰奇是国际自然保护联盟海蛇专家组的成员，是国际自然保护联盟对全球锯鳐现状调查的参与者，同时也是丛林遗产的研究伙伴。她还担任过珊瑚礁生态系统方面的讲师。她管理分子生态和进化学系实验室，并在海洋和热带生物学院从事分子遗传学研究。她是著名的科学大使，从事专业和科学的水肺潜水。布兰奇热衷于应用环保研究，是沿海海豚保护和海洋保护运动中的积极参与者。她目前的研究对象是纯正的澳大利亚西海岸海蛇。

图书在版编目（CIP）数据

冈特生态童书. 第七辑: 全36册: 汉英对照 /
（比）冈特·鲍利著;（哥伦）凯瑟琳娜·巴赫绘;
何家振等译. —上海: 上海远东出版社, 2020
ISBN 978-7-5476-1671-0

Ⅰ.①冈… Ⅱ.①冈… ②凯… ③何… Ⅲ.①生态
环境–环境保护–儿童读物—汉英 Ⅳ.①X171.1-49

中国版本图书馆CIP数据核字（2020）第236911号

策　　划　张　蓉
责任编辑　祁东城
封面设计　魏　来　李　廉

冈特生态童书
水中的食肉植物

[比]冈特·鲍利　著
[哥伦]凯瑟琳娜·巴赫　绘

李原原　译

记得要和身边的小朋友分享环保知识哦！
八喜冰淇淋祝你成为环保小使者！